ICME-13 Topical Surveys

Series editor

Gabriele Kaiser, Faculty of Education, University of Hamburg, Hamburg, Germany

More information about this series at http://www.springer.com/series/14352

Carolyn Kieran · JeongSuk Pang
Deborah Schifter · Swee Fong Ng

Early Algebra

Research into its Nature, its Learning,
its Teaching

 **Springer** Open

Carolyn Kieran
Département de Mathématiques
Université du Québec à Montréal
Montréal, QC
Canada

JeongSuk Pang
Department of Elementary Education
Korea National University of Education
Chungbuk
Korea, Republic of (South Korea)

Deborah Schifter
Education Development Center, Inc.
Waltham, MA
USA

Swee Fong Ng
National Institute of Education
Nanyang Technological University
Singapore
Singapore

ISSN 2366-5947   ISSN 2366-5955   (electronic)
ICME-13 Topical Surveys
ISBN 978-3-319-32257-5   ISBN 978-3-319-32258-2   (eBook)
DOI 10.1007/978-3-319-32258-2

Library of Congress Control Number: 2016935597

Printed on acid-free paper

This Springer imprint is published by Springer Nature
The registered company is Springer International Publishing AG Switzerland

Main Topics You Can Find in This ICME-13 Topical Survey

- A brief history of the early algebra movement and its evolution
- The nature of early algebra, its processes, and its mathematical content
- Research on early algebra learning
- Research on the teaching of early algebra, including the need for professional support for the teacher of early algebra
- A neurocognitive perspective on early algebra.

Contents

Chapter 1
Introduction

This survey of the state of the art on research in early algebra traces the evolution of a relatively new field of research and teaching practice. With its focus on the younger student, aged from about 6 years up to 12 years, this document reveals the nature of the research that has been carried out in early algebra and how it has shaped the growth of the field. The survey attempts to tease out of this evolving and steadily growing research base both the nature of algebraic thinking and the ways in which this thinking can be developed in the primary (elementary) and early middle school student. Mathematical relations, patterns, and arithmetical structures lie at the heart of early algebraic activity, with processes such as noticing, conjecturing, generalizing, representing, justifying, and communicating being central to students' engagement. The role of natural language in the development of early algebraic thinking is considered fundamental. Examples drawn from some of the research in the learning and teaching of early algebra are presented, along with findings from recent neurocognitive studies that offer insights into algebraic thinking and its related activity. This topical survey of the developing field of research study and teaching practice in early algebra should be of interest both to the newcomers to the field, as well as to the more experienced.

© The Author(s) 2016
C. Kieran et al., *Early Algebra*, ICME-13 Topical Surveys,
DOI 10.1007/978-3-319-32258-2_1

Chapter 2
Survey of the State of the Art

This second chapter of the ICME-13 Topical Survey on Early Algebra is divided into five sections: (1) a brief history of the early algebra movement and its research during the first phase of its development in the years leading up to the early 2000s, (2) recent research on early algebra learning and the further evolution of the field since the early 2000s, (3) early algebra in the elementary classroom, (4) a neurocognitive perspective on early algebra, and (5) concluding remarks.

2.1 Brief History of Early Algebra Movement and Its Research up to the Early 2000s

This first section of the chapter deals with the early days of the movement and points to the main areas of research that were springing up in various parts of the world in this new field of early algebra. Up to about the time of the 12th ICMI Study Conference on The Future of the Teaching and Learning of Algebra (Stacey et al. 2004), held in Australia in 2001, the research related to early algebra tended to be rather patchwork in nature, as researchers grappled with questions related to what the content and central focus of early algebra could or should be. During the years that followed, the field of early algebra came to be more clearly delineated; this later historical evolution is sketched in Sect. 2.2.

2.1.1 The Early Algebra Movement

The algebra research carried out during the latter decades of the 20th century with 12- to 15-year-olds pointed to some of the shortcomings of an arithmetic way of thinking when students first experience algebra in high school. The countless

© The Author(s) 2016
C. Kieran et al., *Early Algebra*, ICME-13 Topical Surveys,
DOI 10.1007/978-3-319-32258-2_2

research studies that had disclosed the difficulties involved in moving from an arithmetic to an algebraic form of reasoning (e.g., Kieran 1992; Linchevski 1995; Rojano and Sutherland 2001; Wagner and Kieran 1989) provided a stimulus for exploring whether certain types of algebraic activity, with a focus on what was coming to be generically referred to as *algebraic thinking*, might be accessible to younger students and thereby aid in making the eventual transition to the more formal study of algebra.

However, as Davis (1995) pointed out, research on the question of whether the study of algebra should be spread throughout the primary and secondary curricula had been underway since the 1960s. His ICME-5 report, titled "Algebraic thinking in the early grades" (Davis 1985), was one of the main influences in early discussions of the question of algebra for children from 6 to 12 years of age. Other influences included the research work of mathematics educators who were suggesting alternative ways of conceptualizing the area of school algebra (e.g., Kaput 1998)—much of this work emanating from the Early Algebra Research Group supported by the U.S. Department of Education in the early to mid-1990s (Kaput et al. 2008b), as well as the initiatives of the National Council of Teachers of Mathematics (NCTM 1989, 2000). At the 1987 Research Agenda Conference on Algebra (Wagner and Kieran 1989), one of the areas deemed sorely in need of research attention was that of algebraic thinking.

At the same time that the early algebra movement was beginning in the USA, parallel developments were occurring, for example, in Russian experimental schools and in Chinese primary education; however, as will be seen, descriptions of this activity were not available in English research publications until somewhat later. Reflections of the broader international interest in this emerging field were also indicated by some of the papers presented at the conferences of the European Society for Research in Mathematics Education (e.g., Bolea et al. 1998) and the Psychology of Mathematics Education (PME) (e.g., Steinweg 2001). The year 2001 witnessed for the first time not only a PME Research Forum dedicated to the theme of early algebra (Ainley 2001), but also the designation of one of the thematic groups at the 12th ICMI Study Conference as the *Early Algebra* group (Lins and Kaput 2004). In this Early Algebra group, international participants presented papers and discussed the early algebra research that had been taking place during the years leading up to the Study Conference. Those discussions emphasized that a key characteristic of early algebraic thinking is the expression of generality—a characteristic that figures prominently in the representative samples of research that are highlighted in the sub-section below.

2.1.2   The Development of Algebraic Thinking in the Early Grades: Some Examples

Research related to the development of algebraic thinking in the early grades was such a new domain of study in the late 1980s that it was not treated as a separate category in the 1992 *Handbook of research on mathematics teaching and learning* (Grouws 1992). It was an emerging body of work that was coming to be referred to around the world as Early Algebra. In contrast to the traditional teaching of algebra that usually begins when students are about 12 years of age, the growing body of work on early algebra tended to focus on the 6- to 12-year-old. However, the interest in *algebraic thinking* is not restricted to the young learner; the term *algebraic thinking* has become central to current algebra research involving the older learner as well (see, e.g., Radford 2010; Zazkis and Liljedahl 2002)—*algebraic thinking* having been defined by, for example, Blanton and Kaput (2004) as "a habit of mind that permeates all of mathematics and that involves students' capacity to build, justify, and express conjectures about mathematical structure and relationships" (p. 142).

In addition to the younger age range in this body of work known as Early Algebra was a subtle shift in emphasis from a traditional content-centered characterization of algebra to that of the mathematical reasoning processes and representations that would seem appropriate for young children, as well as to the nature of the early algebra activities that might promote the development of these processes and representations. In particular, the main focal themes during the years leading up to the early 2000s included: (i) generalizing related to patterning activity, (ii) generalizing related to properties of operations and numerical structure, (iii) representing relationships among quantities, and (iv) introducing alphanumeric notation.

2.1.2.1   Generalizing Related to Patterning Activity

Kaput and Blanton (2001) suggested that the *algebrafication* of arithmetic involves moving beyond a proficiency-oriented view to that of developing in the elementary grades the ways of thinking that can support the later learning of algebra. Central to this perspective, according to these two researchers, is the aspect of algebra that includes generalization, and the ways in which this aspect can be capitalized on in the elementary grades. By far, generalizing from numerical and geometric patterns witnessed the largest amount of development and research interest.

An early example of this focus is drawn from an Australian study by Bourke and Stacey (1988) with 371 students, aged 9–11 years of age, on a linear pattern that involved representations of ladders of various lengths. According to the researchers, none of the students had difficulty with finding a way to generalize; but they tended to grab at quick solutions (such as multiplying the number of rungs by 3) and did not subject their responses to critical thinking or to testing them in the face of the given data (see also Stacey 1989). Findings such as these prompted researchers and educators to advocate for a much greater variety in patterning tasks (e.g., Orton 1999), as well as to begin to grapple theoretically with questions on the nature of algebraic thinking and the way it relates to generalization (e.g., Mason 1996; Radford 2000).

Although pattern finding in single-variable situations was becoming fairly common in elementary mathematics curricula, Blanton and Kaput (2004, p. 142) argued that elementary school programs that aim at promoting algebraic reasoning should extend further to include functional thinking (which they defined as "representational thinking that focuses on the relationship between two or more varying quantities"). From the studies they carried out in prekindergarten up to 5th grade, they found that students as young as those in kindergarten could engage in co-variational thinking and 1st graders could describe how quantities corresponded.

2.1.2.2 Generalizing Related to Properties of Operations and Numerical Structure

The thesis underlying the work of the Carpenter et al. (2003) research team was that, if students understand their arithmetic in such a way as to be able to explain and justify the properties they are using as they carry out calculations, they will have learned some critical foundations of algebra. These researchers considered that students not only make sense of the basic operations and procedures within the context of word problems (Carpenter et al. 1999; see also Schifter 1999), but also that such activity serves as occasions to reflect on important properties of these operations. Tasks involving true/false and open number sentences (many drawn from the earlier work of Davis (1964), in the *Madison Project*—e.g., Is $9 + 5 = 0 + 14$ true or false? What is the value of Δ in $18 + 27 = \Delta + 29$?) were found, by the researchers, to be extremely effective. The results of the research of Carpenter et al. (2003) included (a) benchmarks related to students' developing conceptions of the equal sign, (b) a classification of the types of conjectures students make, and (c) a host of rich descriptions of the ways in which young students come to be aware of properties and learn to use them to articulate and justify conjectures. Their research was an influential precursor of several early algebra studies involving number, operations, and properties, which were to follow in the ensuing years.

Another example involving generalizing about number relations was the research of Fujii (2003) and Fujii and Stephens (2001). Fujii introduced young Japanese students to algebraic thinking through generalizable numerical expressions, using numbers as *quasi-variables*—for example, with number sentences such as $78 - 49 + 49 = 78$, which are true whatever number is taken away and then added back. Fujii (2003) claimed that these expressions "allow teachers to build a bridge from existing arithmetic problems to opportunities for thinking algebraically without having to rely on prior knowledge of literal symbolic forms" (p. 62).

2.1.2.3 Representing Relationships Among Quantities

In contrast to much of the numerically-related research on early algebraic thinking, the Russian-based approach developed by Davydov and his colleagues (Davydov et al. 1999) emphasized the teaching of algebra based, not on its numerical foundations, but on relationships among quantities and involving the use of literal symbols right from the first grade. Schmittau translated the three-year Davydov curriculum and implemented it in a USA primary school setting (see Schmittau and Morris 2004). In reference to this curriculum where part-whole relationships are at the core, Schmittau and Morris (2004) have stated that "children write 'if C < P by B, then C = P − B and C + B = P'; the notation indicates that they can move from an inequality to an equality relationship by adding or subtracting the difference, and that addition and subtraction are related actions" (p. 81). They have argued further that this approach "develops theoretical thinking, which according to Vygotsky comprises the essence of algebra" (p. 83). Dougherty (2003), who developed a program in Hawaii based on the approach of Davydov, which she calls *Measure Up*, found that 3rd graders' use of algebraic symbols and diagrams, which evolves within measuring situations, "positively impacts on their mathematical development especially when used ... [within] an approach that simultaneously links the physical model, intermediate representations, and symbolizations" (Dougherty and Slovin 2004, p. 301).

The Russian-based approach does not teach children to solve equations by thinking about "doing and undoing" numerical operations but by direct comparisons between quantities. This is quite different from the Singaporean approach to developing algebraic thinking in the early grades (Ng 2004), where part-whole relationships are also involved, but doing and undoing is considered central. The Singapore elementary mathematics curriculum stresses three thinking processes: analyzing parts and whole, generalizing and specializing, and doing and undoing. An integral part of the curriculum is the *model method* (or pictorial equation, as it is sometimes referred to)—a diagrammatic tool for representing quantitative and numerical relationships and for solving related problems. It is believed that, if children are provided with a means to visualize a problem, they will come to see its structural underpinnings. An example of a word problem that is represented and solved by the *model method* is provided in Fig. 2.1 (see Ng 2004, for details).

While the Singaporean model method does not involve algebraic symbols or methods, the Chinese elementary curriculum for Grade 5 (10- and 11-year-olds) focuses on application problems where students are taught to use both arithmetic and algebraic solving methods (Cai 2004). According to Cai, the aim in teaching younger students both arithmetic and algebraic methods is not only to help them attain an in-depth understanding of quantitative relationships but also to guide them to see the similarities between arithmetic and algebraic approaches and thus to make for a smoother transition from arithmetic to algebraic thinking.

2.1.2.4   Introducing Alphanumeric Notation

As can be surmised from some of the examples presented thus far, the question of whether early algebra should or should not involve symbolic forms was a

Fig. 2.1 Representation and
solving of a problem by the
model method within the
Singapore elementary
mathematics curriculum (Ng
2004)

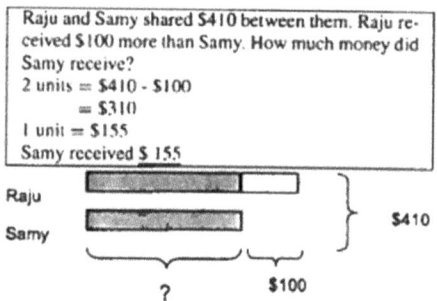

hotly-debated issue during the decades leading up to the early 2000s. Proponents of
an early introduction to algebraic thinking, but without the use of algebraic symbolism, included the researchers who developed the curriculum project,
Investigations in Number Data and Space for kindergarten up to 5th grade (TERC
1998; see Noble et al. 2001). Mathematical change, patterns and relationships,
representation, and modeling were the key focus areas of this project. Moyer et al.
(2004), who used the Driscoll (1999) framework to analyze this project (the
Driscoll framework includes the following algebraic "habits of mind":
doing-undoing, building rules to represent functions, and abstracting from com-
putation), noted that, while students are expected to come up with a general rule, "it
is not the intent of the curriculum that students develop the ability to formally
represent functions with algebraic symbols" (p. 31). In fact, students' use of natural
language to express arithmetical relationships, properties, and generalized pattern
structures was considered by many researchers (e.g., Malara and Navarra 2003;
Radford 2000) to be central to developing and expressing algebraic thinking, as
well as being a mediator in the longer-term construction of alphanumeric modes of
representation.

Even if *Early Algebra* does not mean moving the traditional algebra program
down to the elementary school level, there was a sense among certain researchers
that alphanumeric notation can and should be gradually introduced to early algebra
learners. The research program of, for example, Carraher et al. (2001; Schliemann
et al. 2003) has taken these researchers into the classrooms of 8- to 11-year-olds to
study the ways in which their innovative instruction leads young students to use
algebraic notation to represent problem situations and to interpret the relations
being represented. They have argued that students of this age can develop an
enlarged sense of the equal sign, represent unknown quantities with a letter, represent relations with variables, work with unknowns, write equations, and even
solve letter-symbolic linear equations (Brizuela and Schliemann 2004).

Despite the evidence that some children of this age are able to use alphanumeric
notation, Warren (2003) has urged caution. Her longitudinal study of young

Australian students' understanding of the use of the equal sign found evidence of a certain persistence of narrow views. Warren (2002) also noted the difficulties that 8- and 9-year-olds experience in handling problems with unknowns. Van Ameron (2002) similarly reported from her study of Dutch students that nudging 11-year-olds to use symbolic formulas is not productive, not even if it is done in a tentative and well-considered way.

While technology was not a major component of much of the early-days research in Early Algebra, there were a few exceptions. For example, the research of Ainley (1999) and Ainley et al. (1998) showed that spreadsheets, with their algebra-like notation and graphing facility, can be quite productive with 8- to 11-year-olds as a tool for emergent algebraic reasoning. Sutherland's (1993) research in the ANA Logo project with 11- and 12-year-olds reported students' successful use of variables to express simple mathematical relationships within the context of teacher-developed and teacher-supported tasks in the Logo programming environment.

2.1.3   Concluding Remarks: Early Algebra Research in Years Leading up to Early 2000s

At the ICME-8 conference in Sevilla in 1996, Kieran (1996) proposed a model of algebraic activity that served a few years later as the basis for a definition of algebraic thinking in the early grades—a definition that did not hinge on the use of the letter-symbolic (Kieran 2004):

> *Algebraic thinking in the early grades* involves the development of ways of thinking within activities for which the letter-symbolic could be used as a tool, or alternatively within activities that could be engaged in without using the letter-symbolic at all, for example, analyzing relationships among quantities, noticing structure, studying change, generalizing, problem solving, modeling, justifying, proving, and predicting. (p. 149).

This characterization of algebraic thinking in the early grades synthesized the main thrusts of the various early algebra studies up to the early 2000s, research that was underpinned by and involved the analysis of relationships among quantities, the development of awareness of numerical structure and properties, the study of change in functional situations, generalizing and justifying, and the solving of problems with a focus on relations.

As will be seen in the subsequent sections of this survey chapter, the research that was to develop over the following decade would yield a more cohesive picture of the area of early algebra, as well as a deeper understanding of algebraic thinking, how it develops, and how its development can be supported at the elementary and early middle school levels.

2.2   Recent Research on Early Algebra Learning and Further Evolution of the Field

This second section of the topical survey deals with research on early algebra learning from the early 2000s onward. As will be seen from the various studies exemplified herein, the field of early algebra has gradually come to be more clearly delineated since the 2000s, bringing with it more comprehensive views and theoretical framings of algebraic thinking. At the core of this recent research has been a focus on mathematical relations, patterns, and arithmetical structures, with detailed attention to the reasoning processes used by young students, aged from about 6 to 12 years, as they come to construct these relations, patterns, and structures—processes such as noticing, conjecturing, generalizing, representing, and justifying. Intertwined with the study of the ways in which these processes are engaged in are the two main mathematical content areas of generalized arithmetic (i.e., number/quantity, operations, properties) and functions. Both of these content areas have their proper concepts and objects, including equality/equivalence, co-variation, variable/quasi-variable, expression, equation, diagrams, tables, graphs, and symbols. In sum, over the past 10–15 years, our view of the field of early algebra research is one that has come to be more explicitly characterizable with respect to its central focus, its reasoning processes, its content areas, and its concepts and objects. In line with this evolution, this section of the topical survey is structured according to the following subsections: the nature of early algebraic thinking, its processes, its mathematical content areas, and brief concluding remarks that include implications for future research. Space restrictions limit severely the number of examples that can be presented herein; thus, overall trends are the main focus (see also, e.g., Cai and Knuth 2011; Carraher and Schliemann 2007; Kieran 2011).

2.2.1   The Nature of Early Algebraic Thinking

Blanton et al. (2011) argue that mathematical structure and relationships are central to the practice of early algebra. For Britt and Irwin (2011), early algebraic thinking involves coming to use numbers and words to express arithmetic transformations in general terms. Carraher and Schliemann (2015) characterize early algebraic thinking in terms of basic forms of reasoning that express relations among number or quantities, in particular, functional relations. In these studies and others, mathematical relations, patterns, and arithmetical structures are deemed to be at the heart of early algebraic thinking.

In the development of early algebraic thinking, the role of natural language is of paramount importance (Cusi et al. 2011; Malara and Navarra 2003). Radford's (2011b) detailed analyses of 2nd graders' algebraic thinking take into account not only their use of natural language, but also their spatial descriptions and gestures. Radford (2006) argues further that "using letters does not amount to doing algebra"

(p. 3). He insists that, fundamentally, algebraic thinking is a particular form of reflecting mathematically and that the use of alphanumeric symbols is not necessary; other semiotic representations may also be used.

To conclude this first subsection on the nature of algebraic thinking, we point to Radford's (2014) recent development of a framework for characterizing algebraic thinking, one that involves the following three key notions: (a) *indeterminacy*: unknown numbers are involved in the given problem, (b) *denotation*: the indeterminate numbers are named or symbolized in various ways such as with gestures, words, alphanumeric signs, or some combination of these, and (c) *analyticity*: the indeterminate quantities are treated as if they were known numbers.

2.2.2 Processes of Early Algebraic Thinking

Blanton et al. (2011) focus on the processes of *generalizing, representing, justifying*, and *reasoning with* mathematical structure and relationships. Russell et al. (in press) focus on *noticing, articulating, representing, justifying*, and *contrasting*. Cusi et al. (2011) focus on *verbalizing, translating, arguing, interpreting, predicting*, and *communicating*. However, the bulk of the research on the development of early algebraic thinking focuses on the process of generalizing—a process inherent to all early algebraic activity, just as it is to all mathematical activity (Mason 2005)—with a whole subcategory of the generalizing-oriented research being related to numeric and geometric pattern generalization (e.g., Cooper and Warren 2011; Moss and London McNab 2011).

In line with the significance given to generalization in early algebraic thinking, Kaput et al. (2008a) emphasize that one critical aspect that makes an activity algebraic is deliberate generalization. Even the use of numbers can be qualified as algebraic in so far as its purpose is not on calculation per se but on the representation of a generic example. For instance, when asked to find the number of squares in 'big' figures in the activity of geometric pattern generalization, second graders were able to generate calculation methods or rules across specific instances by using numbers in a general way (Radford 2011b).

Based on his several earlier studies involving patterning activities, Radford (2003) has theorized three layers in students' generalizing activity: (a) *factual generalization* which employs concrete-level actions usually associated with gestures, words, and perceptual activity, (b) *contextual generalization* which rests on situated descriptions of the objects and their naming (such as, referring to the "figure" and the "next figure"; however, these generalizations are not considered algebraic by Radford because algebraic generalizations comprise objects that are non-situated and non-temporal, with no access to a point of reference that involves seeing the objects), and (c) *symbolic generalization* which uses symbols or signs to express the generalization. In 2006, Radford recast his earlier "layers of algebraic generality" into the following definition of *algebraic pattern generalization:*

"Generalizing a pattern algebraically rests on the capability of grasping a commonality noticed on some elements of a sequence S, being aware that this commonality applies to all the terms of S, and being able to use it to provide a direct expression of whatever term of S" (Radford 2006, p. 5).

Rivera (2013) has observed that the process of pattern generalization is not linear, hierarchical, and transitional. Instead it is multidimensional, dynamic, and emergent in character, influenced by cognitive, sociocultural, and other factors. These characteristics are reflected in his *theory of graded representations in pattern generalization* in which various coordinations and connections among different layers (i.e., input, relation, representation, hidden, and output) shape pattern generalization processing. This theory is unique in that it explains both individual differences in performing pattern generalization and continuously evolving characteristics based on learning and experience.

2.2.3 Mathematical Content Areas of Early Algebraic Thinking

Kaput (2008) has stated that three content strands involve algebraic thinking: algebra as the study of structures and relations arising in arithmetic, algebra as the study of functions, and algebra as a cluster of modeling languages. In that most of the early algebra research integrates the third strand within the other two by means of various problem contexts, the main focus here is the first two of Kaput's content strands. It is noted that Kaput's first strand, that of the study of structures and relations arising in arithmetic, has at times been referred to recently as *generalized arithmetic*. In the past, the term *generalized arithmetic* was synonymous with letter-symbolic algebra, with its equations and unknowns. However, within the context of early algebra, *generalized arithmetic* has acquired a much broader sense in that the relations and properties inherent to arithmetical operations are explored and seen by students as being generalizable, without necessarily involving alphanumeric symbols. As will be seen, some of the recent research in early algebra learning that adopts a generalized arithmetic perspective includes work with alphanumeric symbols and some of it does not. Other studies adopt a functional perspective, and yet others combine both perspectives. Section 2.3 of the chapter revisits these first two of Kaput's content strands of early algebra, providing a detailed scenario of each.

2.2.3.1 A Generalized Arithmetic Perspective on Content

In early algebra, generalized arithmetic not only includes number/quantity, operations, properties, equality, and related representations and diagrams, but also can include variables, expressions, and equations—depending on whether or not alphanumeric symbols have been integrated into the learning environment. Some of the recent research in this content area is situated in young students' arithmetic work with number and addition and subtractions operations (e.g., Blanton et al. 2015b), and is extended to include experimentation with students' use of variables to represent unknown quantities. For example, Blanton and her colleagues conducted a teaching experiment in this content area and found that nearly 75 % of the 3rd grade students participating in the intervention program learned to represent unknown quantities with variable notation, even though they had assigned a specific numerical value to the unknown at the outset of the study.

Other studies have been carried out that have probed children's understanding of the equal sign, expressions, and equations. For instance, Matthews et al. (2012) developed a construct map for students' various facets of knowledge of the equal sign in terms of four levels (i.e., *rigid operational, flexible operational, basic relational,* and *comparative relational*). They designed a comprehensive set of tasks with four types of items to assess students' understanding of the equal sign and ultimately of mathematical equality. The tasks were given to 224 students in Grades 2–6. Results indicated that students were sensitive to the equation formats as well as the location of the operations. Providing verbal explanations for the advanced relational reasoning items remained challenging even for upper grade students. A noticeable result was that the children with advanced understanding of the equal sign tended to solve difficult equations, which suggests a direct link between knowledge of the equal sign and algebraic thinking.

In a New Zealand project involving an intervention program promoting early algebraic thinking and that included comparative studies with students in a typical arithmetic-based curriculum, Britt and Irwin (2011) found that students using the new curriculum developed by the project were more successful than their counterparts using the conventional curriculum in solving test items—items that included not only simple compensation in addition but also complex equivalence with fractional values. With an additional longitudinal study including students aged 12–14, the researchers demonstrated that early and sustained exposure to algebraic thinking in elementary school leads to more sophisticated generalization involving the alphanumeric symbols of algebra in intermediate school.

2.2.3.2   A Functional Perspective on Content

Within a functional perspective on the mathematical content of early algebra, the concept of co-variation and its related notion of change are central, as are representations such as tables, graphs, and other such function-oriented diagrams. The objects of variable, expression, and equation are also involved, but with a different interpretation from that held within the perspective of generalized arithmetic. Blanton et al. (2011) argue that functional thinking entails "generalizing relationships between co-varying quantities, expressing those relationships in words, symbols, tables, or graphs, and reasoning with these various representations to analyze function behavior" (p. 13).

A large number of past studies of students' functional thinking have focused on the upper primary and middle school grade students (e.g., Ellis 2007). However, some researchers investigating students' understanding of functional relationships suggest that even younger children, with age-appropriate pedagogical support, are able to engage in co-variational thinking and can represent how two varying quantities correspond in multiple ways, including with letters as variables. Moss and London McNab (2011) have reported that second graders could figure out a general function rule by focusing on the relation between the position number and the number of blocks in geometric growing pattern activities. The students were also able to develop a robust understanding of two-part function rules (i.e., $y = ax + b$) by noticing the constant in visual arrays and representing it in their natural language.

Blanton et al. (2015a) have emphasized that even some of the first graders in their study were able to generalize various functional relationships between co-varying quantities and that their initial levels of understanding could become further sophisticated with the support of well-designed instruction. Within that same study, Blanton and her colleagues developed a learning trajectory to describe first-grade children's (6-year-olds) thinking about generalizing functional relationships. The trajectory, which can serve as a framework for related work on the development of young children's functional thinking, involves different levels of sophistication in generalizing functional relationships by specifying whether children can (a) notice mathematical features in a task, (b) understand the relationships between quantities through recursive thinking or functional thinking, (c) observe the regularity within particular instances or otherwise across all instances, (d) describe a functional relationship in a generalized form, (e) elaborate on two quantities being compared and the functional relationship between them, and (f) deal with function as an object while understanding the boundaries of the generality. Understanding the characteristics of such levels is significant because it sheds light on how sophisticatedly young children condense the functional relationships.

From their longitudinal studies of 3rd to 5th graders, Carraher and Schliemann (2015) have found that students as young as 8–9 years of age can use relations to derive other relations. For example, with the statement, "Tom is 4 inches taller than Mary and Mary is 6 inches shorter than Leslie," students were able to derive Tom's height from Mary's and Leslie's heights, to derive Mary's height from Tom's and Leslie's heights, and to derive Leslie's height from Tom's and Mary's heights. They learned to express this ternary relation on a number line with a "variable origin N"—what the researchers termed the "N-number line," and in which positions were denoted as N-2, N-1, N, N+1, N+2, and so on.

While some of the past research involving younger students' algebraic thinking has focused on the challenges of introducing functional ideas, such as for example, shifting students' focus from a recursive to an explicit functional perspective (Warren and Cooper 2008), the more recent work suggests that students in the lower primary grades, even in kindergarten, may be far more able to begin to think algebraically than was previously imagined. But this does not occur spontaneously.

It requires, in the words of Bass and Ball (2003), "supports for the mathematical work of the teacher in pressing students, provoking, supporting, pointing, and attending with care" (p. vii).

2.2.4   Implications for Future Research

We have attempted in this section of the chapter to illustrate the main trends in research since the early 2000s on the development of algebraic thinking in early algebra learners, and at the same time offer our perspective on how the field has come to be more clearly delineated—in terms of the nature of algebraic thinking, its reasoning processes, and its mathematical content areas. In these concluding remarks, we wish to draw out a few implications for future research. Our first comment concerns the developmental aspect of algebraic thinking. Some studies, specifically those that compare an intervention group with a non-intervention group, show clearly that algebraic thinking is not naturally developed through traditional arithmetic-based instruction and curricula as students progress through elementary and middle school. In order to develop algebraic thinking, essential ways of thinking algebraically need to be intentionally fostered in instruction from the earliest grades. There is a growing body of research that provides empirical evidence of how the development of early algebraic thinking evolves into more sophisticated ways of thinking. In turn, these more sophisticated ways of thinking serve to influence subsequent learning of important algebraic concepts given longitudinal interventions. Nevertheless, more systematic and long-term investigation is needed to show the impact of early algebraic thinking on the later study of algebra.

Another area in need of further exploration and study is the development and use of digital tools in early algebra research within the content area of number, operations, and properties. Much of the recent research integrating technology in the learning of algebraic thinking has been conducted in the areas of pattern generalization and functional thinking, and with students in the 12- to 14-year-old age range (e.g., Mavrikis et al. 2013; Roschelle et al. 2010). While there have been some exceptions, such as the research of Hewitt (2014) with 9- and 10-year-olds in the *Grid Algebra* environment, which has been shown to extend students' understanding of numerical operations, few other digital environments have been designed with the aim of developing primary school students' algebraic thinking in this area.

One final comment arising from our survey of recent studies—one that is related to the previous remarks on the use of digital environments in early algebra learning—concerns theoretical development. Much of the theory development that has occurred has been in conjunction with empirical investigations focusing on the areas of algebraic pattern generalization and functional thinking. However, little theorization has taken place regarding the area of number, operations, and properties, even though this area is one of the main critical routes to fostering early algebraic thinking.

2.3 Bringing Early Algebra into Elementary Classrooms

This third section of the chapter addresses the following questions: What is the nature of early algebraic content in classroom contexts? What interactions among teacher and students sustain students' engagement with early algebra? What supports are necessary?

2.3.1 The Nature of Early Algebraic Content in Classroom Contexts

To begin an examination of early algebra in classroom settings, consider two scenes, the first involving algebra as the study of functions, the second, as the study of structures arising in arithmetic and quantitative reasoning.

In the first scene, a fourth-grade Canadian class was presented with the following situation: *For his birthday, Marc received a piggy bank with one dollar. He saves 2 dollars each week. At the end of the first week he has 3 dollars; at the end of the second week he has 5 dollars, and so on.* Students were given red and blue chips to model the situation for Weeks 1–5 and were asked to figure out how much Marc would have after Weeks 10, 15, and 25. Krysta and Albert's model for the first 5 weeks looks like the display in Fig. 2.2. (The top chip in each configuration is blue; the rest are red.) Now they are discussing the amount of money saved after 10 weeks.

Krysta   So we should do… That (pointing to the chips for week 5) times two. So 11.
Albert   11 plus 11. 22.
Krysta   22.
Albert   Well, wait. No. It would be 11 plus 10 because (pointing to the blue chip). We always start with the [blue chip] (Radford and Roth 2011, p. 235).

Like much algebraic work in elementary classrooms, these students are engaged in the study of functions (e.g., Blanton 2008; Carraher et al. 2006; Malara and Navarra 2003; Moss and London McNab 2011)—the amount of money in Marc's piggy bank is a function of the number of weeks that have elapsed. We see two students working together to find the value of the function at particular points. Krysta has made a common error, assuming that the value at 10 weeks should be double the value at 5 weeks. But with chips to represent Weeks 1–5 laid out before them, Albert recognizes what is the same across the weeks and realizes that they have to pay attention to that blue chip, which they always counted first.

Fig. 2.2 Krysta and Albert's model

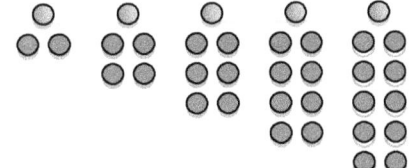

What is important in this lesson is not simply that the students extend the sequence to answer the stated problems. Students could lay out chips and count the number of chips without engaging in algebraic thinking. Nor is it about coming up with a rule that defines the amount of money in a piggy bank as a function of week number. Students might guess and check, testing a sequence of rules—$n + 1$, $n + 2$, $n + 3$—until they find one that fits the values they have determined.

Rather, as we begin to see with Albert, it is recognizing what is general across problems. The aim is for students to see their models not merely as a bunch of chips, but as a collection that can be decomposed, to see how the term number is related to the decomposed parts, and to see how the number of chips for each week can be calculated from the term number (Radford and Roth 2011). The work of noticing the underlying structure to generalize across specific instances is what makes this an example of algebraic thinking.

In the second scene, a fourth-grade USA class is investigating what happens to the product of a multiplication expression when one factor is increased by a given amount. They have already discussed what happens when one factor is increased by 1 or by 5, and now they consider what happens when a factor increases by 2. Prior to this lesson, students were shown pairs of equations such as those shown in Fig. 2.3 and were given the prompt, "When I add 2 to a factor, then this happens to the product."

The lesson opens when the teacher presents four students' statements.

1. When I add 2 to a factor, it changes by the number of groups.
2. For $7 \times 3 = 21$ and $9 \times 3 = 27$, the product changes by two 3s. And for $7 \times 3 = 21$ and $7 \times 5 = 35$, the product increases by two 7s.
3. When I add 2 to a factor, I take the other factor and multiply it by 2.
4. When I add 2 to a factor, the product increases by 2 groups of the other factor.

The class comments on each statement, working to understand its meaning. During the discussion, the difference between 7×3 and 9×3 is illustrated by drawings like those shown in Fig. 2.4.

Fig. 2.3 Related equations prompt an examination of structure

$7 \times 3 = 21$	$7 \times 3 = 21$
$7 \times 5 = 35$	$9 \times 3 = 27$

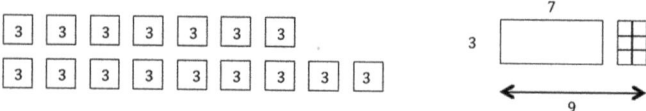

Fig. 2.4 Representations to compare 7 × 3 and 9 × 3

Throughout the discussion, the teacher asks students to show how the representations illustrate the expressions, describe the correspondences across representations, and explain why their ideas work for any multiplication expression, not just 7 × 3. Then she asks the class to extend their claim.

Teacher We've been talking about adding 2 to a factor, and last week we talked about adding 5. Could we come up with a conjecture that would work for any number added to a factor? Does it work only because we're adding 2?

Anita If you add any number to one of the factors, then the other factor increases by that number.

Several students disagree with Anita's statement.

Teacher Let me write it so we can see if we all agree or disagree.

Anita If any number gets added to one of the factors, then the product gets increased by the other factor times the number that was added to the factor.

Kevin This should be our class conjecture.

Teacher Megan, you want to add on to that or say something different?

Megan It's kind of the same, sort of like Lila's up there. When I add any number to a factor, I take the other factor and multiply it by the number that I added onto one of the factors (Russell et al. in press).

These students are engaged in algebra as the study of structures in arithmetic and quantitative reasoning (e.g., Britt and Irwin 2011; Russell et al. 2011; Schifter et al. 2008a). In different studies, researchers use various strategies to focus students on structures of the operations. Carpenter et al. (2003) ask students to evaluate number sentences as true or false to bring their attention to structure. In the Davydov-based *Measure Up* curriculum (Dougherty 2008), students compare continuous quantities—area, length, volume, or mass—without reference to numbers. In the scene above, students look for patterns in related equations. In each of these settings, students learn to notice regularity, to articulate generalizations, and to explain or prove their conjectures.

Among the goals of these lessons is that students learn the language of generalization. Yet, in the classroom presented above, the objective is not to provide students with the most precise statement of the distributive property. Rather, students use their own language and work together to create a statement that is clear enough for someone outside the class to understand. This is one phase of what the

Italian researchers Malara and Navarra (2003) call "algebraic babbling." Analogous to the way children learn natural language, students learn to communicate in algebraic language by starting from its meaning, and through collective discussion, verbalization, and argumentation, gradually become proficient in syntax. These fourth-graders will go on to represent the phenomenon they are articulating with story contexts, diagrams, and pictures of groups in order to make the case for their conjecture.

The study of the behavior of the operations helps students come to see an operation not exclusively as a process or algorithm, but also as a mathematical object in its own right (Sfard 1991; Slavit 1999). However, Russell and her colleagues found that, once students noticed regularity in calculations and articulated conjectures, many assumed that same regularity applied to all operations (Russell et al. 2011). For example, students may be explicit about a rule for creating equivalent addition expressions—if you add some amount to one addend and subtract that same amount from the other, the sum remains the same—and believe that the same pattern applies to subtraction or multiplication. At the end of a sequence of lessons that explored equivalent addition expressions and equivalent subtraction expressions, as the class reflected on their work, one third-grader said, "When we got the idea of seeing if our addition rule works for subtraction, I was like, of course it works. And then it was like uh-oh, it doesn't work, and I lost all hope. I'm happy we found a very close but different rule" (Russell et al. in press). For this reason, it is important that students contrast the behavior of different operations.

Although the two scenes presented above illustrate different aspects of algebra, there is much in common. Most prominently, students are thinking analytically about indeterminate numbers (Radford 2014). They are looking for structure, whether in a function or in the behavior of an operation, the key to algebraic reasoning.

In both cases, students link spatial and numerical representations of structure. In the language of Radford (2011a), "the awareness of these structures and their coordination entail a complex relationship between speech, forms of visualization and imagination, gesture, and activity on signs (e.g., number and proto-algebraic notations)" (p. 23). Warren and Cooper (2009) hypothesize that "abstraction is facilitated by comparing different representations of the same mental model to identify commonalities that encompass the kernel of the mental model" (p. 90). Moss and London McNab (2011) theorize that "the merging of the numerical and the visual provides the students with a new set of powerful insights that can underpin not only the early learning of a new mathematical domain but subsequent learning as well" (p. 280).

In neither of these lessons are students engaged in algebraic notation. Although different researchers emphasize conventional notation to different degrees, they converge on the notion that algebraic notation does not necessarily indicate algebraic thinking, and algebraic thinking does not necessarily entail conventional use of letters (Britt and Irwin 2011; Radford 2011a, 2014). Students might use pictures, gestures, or natural language to communicate a generalization. They might point to a stack of cubes and say, "This can be any number." At times, they might communicate their generalization with the use of specific numbers—"These are equal groups and they could each be a million; if you add 1 to a factor, you add another group, and the product goes up by a million"—what Mason (1996) calls "seeing the general through the particular."

2.3.2   Roles of Students and Teachers in Classrooms

The goal of early algebra is to promote a way of thinking—the habit of looking for regularity, and articulating, testing, and proving rules or conjectures for an infinite class of numbers. This is achieved through classroom interaction around ideas, sometimes in pairs and small groups, but largely through class discussion in which students elaborate their own thinking and engage with their classmates' ideas. Together they consider, evaluate, challenge, and justify hypotheses. Students contribute different pieces of information and build upon others' explanations to jointly create a complete idea or solution. Over the last two decades, there has been a growing body of work to study the impact of such discussion on learning outcomes. Empirical findings support the hypothesized benefits of active student participation in discussion (Webb et al. 2014).

In order to establish a setting in which students engage in this manner, a teacher must first set the expectation that students support their ideas with explanations and probe and challenge each other's ideas to make sure they follow classmates' reasoning (Blanton and Kaput 2008). Teachers ask probing and clarifying questions to help students make the details of their thinking explicit. They acknowledge and validate students' proposals to encourage sustained discussion and help students confront discrepancies in their thinking. At times, the teacher may offer suggestions to help students consider and develop new options. The teacher directs the discussion, filtering students' ideas to draw their attention to what the teacher determines is pertinent and meaningful (see e.g., Cusi et al. 2011; Kazemi and Stipek 2001).

Teachers attend to correct and well-formulated ideas as well as ideas that are in development or even incorrect. For example, in the first scene above, Krysta's incorrect idea—double the amount of money after 5 weeks to find the amount of money after 10 weeks—is worthy of attention. The teacher can help Krysta, and perhaps the whole class, examine the result of doubling the components of the problem to understand *why* her strategy does not work for this function. In the second scene, it would have been useful to discuss the error in Anita's first statement had she not changed it.

Although many of these aspects of classroom interaction may be relevant to any topic of discussion, especially applicable to early algebra is to involve students in metacognitive acts (Cusi et al. 2011), to reflect on their own observations to move to a level of generalization and argument. In the language of Malara and Navarra (2003), students "substitute the act of calculating with looking at oneself while calculating" (p. 9). Franke et al. (2008) note that instead of asking, "How did you solve the problem?" (which teachers often ask when the class is working on calculation or solving story problems), in an early algebra lesson teachers ask, "How did you know that?" "Will that work for all numbers?" or even, "What is it that will work for all numbers?"

The teachers in Russell et al.'s (in press) project reported that, when working on early algebraic topics, they noticed a feature of their teaching that they named *productive lingering* (Russell 2015). When the topic of discussion is complex and abstract with the opportunity to make many connections, even after an idea has been clearly stated and the class seems to be in agreement, if the teacher asks another question, the class continues to engage, offering new insights, taking the discussion deeper. A glimpse of productive lingering is offered in Scene 2. Anita has offered a correct statement, which Kevin declares could be the class conjecture, but the teacher provides space for further discussion. Megan offers a different formulation, taking on one of her classmates' statements from the beginning of the lesson, but changing the language just enough to extend it from adding 2 to a factor to adding *any number* to a factor. However Megan's formulation is not complete. There is still more for the class to consider.

2.3.3 What Can Happen in Classrooms in General?

Several longitudinal studies have demonstrated positive learning outcomes from early algebra instruction (Blanton et al. 2015b; Radford 2014; Warren and Cooper 2009). However, in most of these studies—the New Zealand Numeracy Project (Britt and Irwin 2011) and the Italian ArAl project (Malara and Navarra 2003; Cusi and Malara 2013) are among the few exceptions—lessons were taught or co-taught by researchers. The next question is, what supports are needed to bring early algebra into classes taught by elementary teachers?

Teachers need curricular materials. Over the last fifteen years, a number of curricular programs have intentionally infused early algebra throughout the grades (Britt and Irwin 2011; Dougherty 2008; Goldenberg and Shteingold 2008; Schifter et al. 2008b). Some researchers have described how early algebra appears in the Japanese (Watanabe 2011) and Singaporean and Chinese (Cai et al. 2011) curricula. Other projects have published materials for teachers to supplement their regular programs (Carpenter et al. 2003; Russell et al. in press).

However, research on teaching and curriculum has revealed that there is a substantial difference between curriculum as written and curriculum as enacted in the classroom. Among the factors that transform written curriculum into enacted curriculum are teachers' beliefs and knowledge, their orientation to the curriculum, their own professional identities, and the school and classroom culture (Stein et al. 2007). Given these factors, prepared curricula may have limited impact on what happens in the classroom. Movement of early algebra into elementary classrooms requires professional development (Blanton and Kaput 2008; Dougherty 2008).

Professional development programs in early algebra share a number of features. First, teachers must learn the content they intend to teach. If teachers understand mathematics as procedures for calculating and solving problems, they must widen their view to include looking for and examining structure. They must understand that this is not a body of content "to master," but that they and their students will continue to discover new connections through the act of teaching (Blanton and Kaput 2008; Dougherty 2008; Franke et al. 2008; Schifter et al. 2008a).

Second, especially if teachers have had no experience in systematically examining student thinking, they must reorient their practice to develop the disposition of listening for students' mathematical ideas and learn to situate those ideas in relation to the content on which the class is working.

Third, teachers must learn how to lead discussions. They must learn how to analyze their students' ideas in the moment and to make judgments about which ideas to pick up on. They must learn types of questions and responses that will draw students' attention to the content to be explored and help them make new connections.

Several early algebra professional development programs report anecdotal evidence of success (Blanton 2008; Cusi et al. 2011; Dougherty 2008; Franke et al. 2008; Schifter et al. 2008a). Britt and Irwin (2011) and Russell et al. (submitted) offer quantitative results of student learning.

2.3.4 Conclusion

As illustrated by the research described in this third section of the topical survey chapter, early algebra offers the promise of not only preparing students for their algebra courses to come, but also deepening their understanding of the properties of the number systems in which they are learning to calculate and instilling habits of such mathematical practices as looking for structure and expressing regularity. Research has demonstrated that, with teachers who are prepared in terms of content and pedagogical practice, young students engage with this content in the context of their classrooms, resulting in positive achievement outcomes. However, to enact early algebra programs on a large scale requires investment in both curriculum materials and long-term professional development.

2.4 A Neurocognitive Perspective on Early Algebra

In this fourth section of the chapter, we highlight recent research in mathematics educational neuroscience that is directly related to algebraic thinking—research that offers new insights into the representations and solving methods associated with algebraic activity and which can thereby serve to inform the field of early algebra research.

Research shows that many Singapore secondary students, instead of using letter-symbolic algebra to solve the algebra word problems found in secondary texts, continue to use the problem solving method of drawing a diagram, known locally as the model method (Khng and Lee 2009). A possible reason for this behavior is that algebra is perceived to be the more abstract of the two methods. Findings from two neuroimaging studies using functional Magnetic Resonance Imaging (fMRI) provided evidence to support this perception. A theoretical framework is used to hypothesize why this is so. Drawing from cross-sectional and developmental work in arithmetic, we caution against using data with adult participants to generalize on how primary pupils' brains may respond and on how they learn to use the model method to (a) represent quantitative information and (b) solve arithmetic word problems and algebraic word problems found in primary texts. Researchers need to be cognizant of developmental issues related to how knowledge of the model method can support algebra novices in learning to use formal algebra to solve the algebra word problems that are found in secondary texts. Neuroimaging work with arithmetic could serve as a signpost to guide future research in algebraic reasoning.

2.4.1   Singapore Model Method to Solve Arithmetic and Algebra Problems

Although many problem solving heuristics are taught at primary level (Curriculum Planning and Development Division 2006), the model method (Ng and Lee 2009) has the greatest currency. The affordance of the model method means it can be used to solve arithmetic problems (top third of Fig. 2.5) and algebra word problems (middle third of Fig. 2.5) that normally require the construction of linear equations up to two unknowns (the algebraic method being shown in the bottom third of Fig. 2.5). The arithmetic word problem that is illustrated in the figure is: *Mary has 12 red counters and 4 times as many blue counters. How many counters has Mary altogether?* The algebra word problem that is illustrated in the same figure is: *A school bought some mathematics books and four times as many science books. The cost of a mathematics book was $12 while a science book cost $8. Altogether the school spent $528. How many science books did the school buy?*

2.4.2   Different Methods Used to Solve Secondary Algebra Word Problems

Although it is acceptable to use alternative methods to solve algebra word problems, students' continued use of these alternative problem-solving strategies may not serve them well in the long term. It is vital for students to acquire a sound

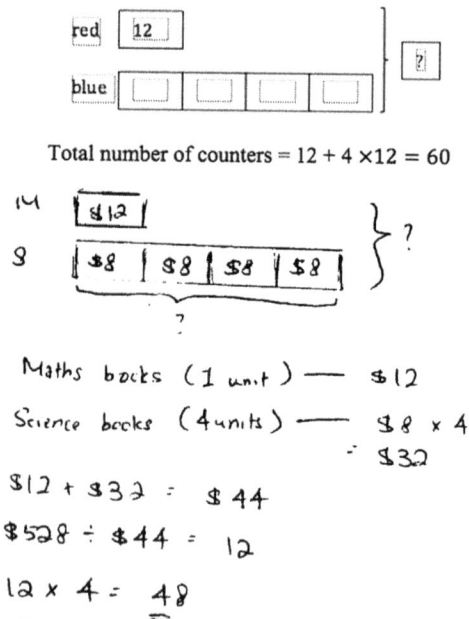

Total number of counters = 12 + 4 ×12 = 60

Maths books (1 unit) —— $12
Science books (4 units) —— $8 × 4
                            = $32

$12 + $32 = $44
$528 ÷ $44 = 12
12 × 4 = 48

Let m be the number of mathematics textbooks and s the number of science books.

$$s = 4m$$
$$12m + 8s = 528$$
$$12m + 32m = 528$$
$$m = 12$$
$$s = 4 \times 12$$
$$s = 48$$

Fig. 2.5 An arithmetic word problem and an algebra word problem, both solved by means of the model method, followed by the algebraic method for the algebra problem

knowledge of algebra and the related procedures and skills should they wish to engage in higher mathematics and other disciplines such as the sciences and economics. In Singapore, many secondary students continue to use problem-solving heuristics, in particular the model method to solve algebra word problems, which can create difficulties for them as they try to coordinate their old method with the new one being taught (Khng and Lee 2009). In one study (Ng 2012), 124 Secondary 2 (14+) students from five schools participated; they were given an hour to respond to 10 algebra word problems. Solutions to two problems are discussed here (see Table 2.1 and Fig. 2.6).

Figure 2.6 shows how (a) the algebraic method, (b) a semi-algebraic method, and (c) the problem solving heuristics taught in Singaporean primary school were used

Table 2.1 Word problems presented to secondary 2 students and the related success rates as a function of algebra (A) versus the model method (MM)

The algebra word problems		A (C)	A (IC)	MM (C)	MM (IC)
1	Parade problem: There are 900 people at a parade. There are 40 more men than women. There are twice as many children as there are men. How many children are there?	39 52[a]	36 48	11 26	32 74
4	Spending problem: Ahmad has four times as much money as Betty. After Ahmad spent $160 and Betty spent $40, they each had equal amounts of money. How much money did Ahmad have at first?	59 81	14 19	13 30	30 70

Note C and IC represent correct and incorrect responses respectively.
[a]Represents percent who used the method. The total is not 100 % because not all methods are recorded. Problems are named to facilitate discussion.

to solve these problems. Any of the unknowns or generators can be used to solve an algebra word problem. The algebra solution to the Parade Problem was found by using the number of children as the generator and representing this unknown with the letter x. The semi-algebraic method involved a combination of the model method and algebra. The letter x in each rectangle suggests that the student knew that the rectangle represented the unknown x, but "doing-undoing" was used to find its value. Perhaps the student thought that the solution was algebraic because the letter x was used to announce the answer. The solution illustrated in the far right panel shows how the model method was used. In both of the latter two methods, the unknown number of women was the generator.

The model drawing for the Spending Problem correctly depicted the amount held by Ahmad and Betty. Instead of constructing a system of equivalent equations to find the value of x, the amount of money held by Ahmad was found by a series of doing-undoing processes. To the left of the model drawing, the letter x was used in representing the amount of money held by each friend; however, it was never featured again in the solving process.

2.4.3   Neuroimaging, the Model Method, and Algebra

Two studies investigated the nature of brain responses when young adult participants were asked to use the model method or algebra to represent and then to solve algebra word problems. Lee et al. (2007) focused on the initial stages of problem solving: that of translating textual information into either a model representation or an algebraic equation. Eighteen adults who were proficient and competent users of both the algebraic and the model methods were presented with algebra word problems. They were then asked to represent the textual information either as algebraic equations or model representations and then validated whether the

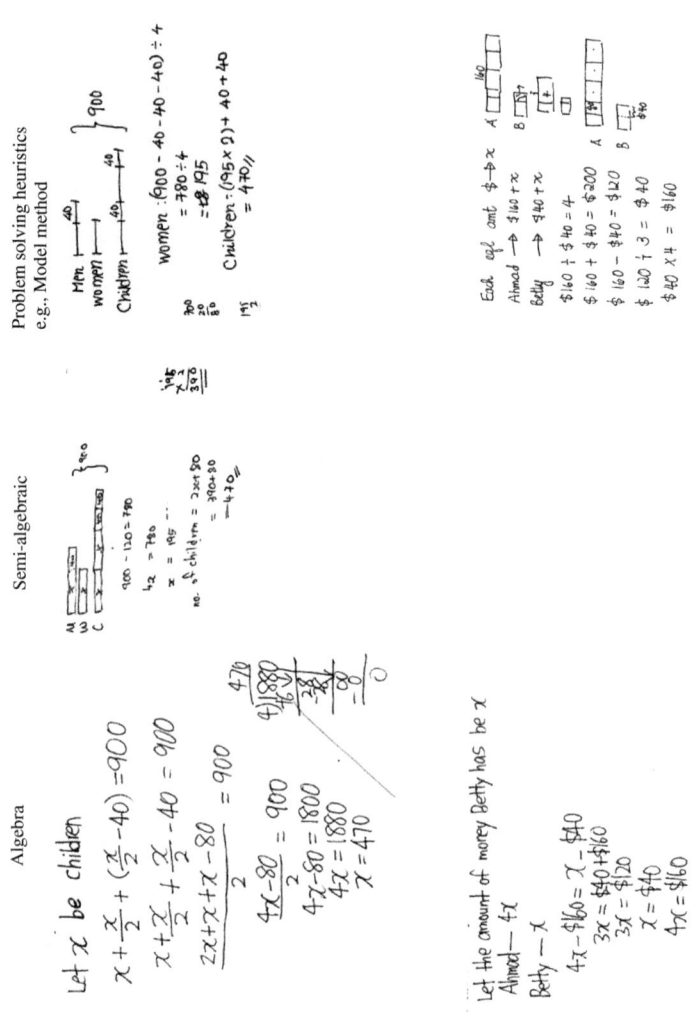

presented solutions matched their representations. In a later study, Lee et al. (2010) focused on the solution phase of the problem solving chain. Here 17 young adult participants were asked to find the value of the unknown when the information was presented either as a model representation or as an algebraic equation. Although both studies showed that the two representations were comparable in terms of area of brain activation in that both activated areas linked to working memory and quantitative processing, algebra imposed greater attentional demands. Evaluating for the unknown using the algebraic equations required "greater general cognitive and numeric processing resources" (Lee et al. 2010, p. 591). For ethical reasons no studies were conducted with Singapore primary school children. The assumption was that children and adults may share similar responses.

2.4.4   Why Algebra May Be the More Resource Intensive of the Two Methods

The following theoretical framework (Ng 2012) constitutes a basis for understanding why algebra could be more resource intensive than the model method. With the model method, the value represented by the rectangle that stands for the unknown can be found by applying the doing and undoing process. With algebra, information in the text is translated into an algebraic equation. The value of the unknown is evaluated by the construction of a series of equivalent equations. Thus the transition from the model method to algebra requires that students know that the role of the rectangle is taken over by the letter.

2.4.4.1   The Transition to Letters as Unknowns Requires an Expansion of the Knowledge Related to the Use of Letters

Letters have different meanings in different situations (e.g., Booth 1984; Kieran 1989; Küchemann 1981; Usiskin 1988). Solving for the unknown x in any equation may require first simplifying the algebraic expressions on either side of the equal sign and then transforming the equations containing these simplified expressions into simpler equivalent equations. Such knowledge construction is no mean feat as it requires developing new cognitive structures and connections associated with operating on letters.

2.4.4.2   Algebraic Expressions Are Legitimate Forms of Answers

When the model method is used to solve algebraic word problems, the information captured by the model is translated into arithmetic expressions, which can be evaluated and a single numerical answer results from operating on the arithmetic expression. For example, the number of children participating in the parade was found by first writing down this arithmetic expression: $(195 \times 2) + 40 + 40$ whose resulting sum is 470. With algebra, however, students must be able to accept algebraic expressions as legitimate answers (Collis 1975; Davis 1975). The algebraic expression $x + x/2 + (x/2 - 40)$ actually represents the number of children at the parade. Because the operators are still visible, many novice learners of algebra experience a product-process dilemma (Davis 1975); they do not treat this algebraic expression as a legitimate answer. Students have to learn to accept that the algebraic expression $x + x/2 + (x/2 - 40)$ does represent the number of children at the parade, and that this expression is a single entity. Early work in algebra begins with students being able to accept that there is a lack of closure in algebraic forms (Collis 1975).

2.4.4.3 Algebraic Representations no Longer Adhere to the Same Set of Conventions Underpinning the Use of Numbers

In arithmetic, operating on whole numbers can be expressed as a single entity as illustrated by this arithmetic equation: $(195 \times 2) + 40 + 40 = 470$. To complicate matters further, the sum of two fractions can be expressed as the concatenation of two numbers. For example, the sum of $2 + \frac{1}{3}$ can be written side by side: $2\frac{1}{3}$. There is no one-to-one correspondence between how numbers can be expressed and how algebraic expressions can be simplified. For example, with algebra, $x + \frac{x}{2}$ cannot be simplified to $x\frac{x}{2}$. Thus the shift from the model method to algebra as a problem solving tool requires construction of new knowledge, namely that algebraic representations no longer adhere to the same set of conventions underpinning the use of numbers (Kieran 1989).

2.4.4.4 Knowledge of Equality-Equivalence of Algebraic Expressions Is Crucial

The shift from the model approach to the use of algebra as a problem solving tool requires construction of knowledge that an algebraic equation is (a) a structure that links two different algebraic expressions and (b) that two or more different algebraic expressions can be constructed to represent the same situation. This equality-equivalence aspect of algebra (see Kieran 1989, 1997) requires sound knowledge of the properties of equality: the reflexive (the same is equal to the same), the symmetric (equality of the left and right sides of each equation), and the transitive (if $a = b$ and $b = c$, then $a = c$).

In arithmetic, the equal sign tends to be treated as a procedural symbol that announces the answer after a series of operations has been conducted (Kieran 1981; MacGregor and Stacey 1999; Stacey and MacGregor 1997). For example, when the model method was used to solve for the amount of money held by Ahmad, the equal sign at the end of the arithmetic expression \$160–\$40 was used to announce the answer (i.e., the difference of \$160 and \$40 is \$120). Contrast that meaning of the equal sign with the one required when algebra is used to solve the same problem. Research (Kieran 1997; Sfard and Linchevski 1994) shows that there are two main sets of conceptual demands associated with solving equations: simplifying expressions and working with equality-equivalence. The cognitive demands needed to simplify expressions have already been discussed in relation to the acceptance of algebraic expressions as legitimate answers; therefore, the focus here is the requirement that solving algebraic equations involves two significantly different conceptualizations of equality-equivalence (Kieran 1997).

Reflexive-equivalence (equality of the left and right sides of each equation): In a conditional equation, for a specific value of x, the resulting values of the expressions on either side of the equal sign are the same. For example, in the Spending Problem the equation $4x - 160 = x - 40$, the expression $4x - 160$ is equal to the expression $x - 40$ when x has the value of 40.

Equivalence of successive equations in the system of equations constructed to solve the problem: To solve a given equation, the conventional procedure is to construct the vertical chain of equivalent equations that will result in the resolution of the unknown value. The equivalence is achieved in one of two ways. First, equivalence is maintained by replacing an expression with an equivalent expression. In the Parade Problem, the expression $x + \frac{x}{2} + \left(\frac{x}{2} - 40\right)$ in the equation $x + \frac{x}{2} + \left(\frac{x}{2} - 40\right) = 900$ was transformed into the expression $(4x - 80)/2$ and was replaced by the latter to yield the subsequent equation $\frac{4x - 80}{2} = 900$ in the equation solving chain. Second, equivalence is achieved by replacing an equation with an equivalent equation (i.e., one having the same solution as the previous equation in the chain) without having to replace the preceding expression with another equivalent expression. The equation $\frac{4x - 80}{2} = 900$ is equivalent to the equation $\left(\frac{4x - 80}{2}\right) \times 2 = 900 \times 2$. In this case equivalence of the previous equation is maintained by multiplying both sides of the equation by the same amount. The expression $\left(\frac{4x - 80}{2}\right)$ is not replaced by any other expression.

Comparing and contrasting the solution of the Parade Problem using algebra against the model method suggests that major cognitive adjustments—'accommodations' rather than assimilations—are needed to solve algebraic equations. Such major accommodations may help explain why neuroimaging studies have found that, of the two methods, algebra tended to activate the procedural part of the brain and it required more attentional resources. In order to maintain equality-equivalence of the series of equations in the equation solving chain, appropriate rules and procedures specific to operating on letters were operationalized. These rules and procedures were no longer identical to those used to operate on numbers. With the model method, solving for the unknown values involved working only with numbers; thus there was no conflict in the conventions used and hence may have been less procedurally driven. In the model method, solving for the unknown involves the processes of doing and undoing. With algebra, solving for the unknown requires using forward operations and mandates the maintenance of equivalence of each single algebraic equation.

2.4.5 Competent Adults and Children Process Arithmetic Information Differently

Competency in arithmetic is fundamental to work related to algebraic reasoning (English and Warren 1998; Lannin et al. 2006). Development of arithmetic skills relies on cognitive processes such as working memory, memory encoding and retrieval, decision-making, and retrieval. The development of such processes is complex and significant changes occur in developing children's brains (e.g., Kwon et al. 2002), thereby making available additional processing resources that enable the more efficient processing of complex cognitive operations (Kail and Park 1994).

Neuroimaging studies conducted with adults, and cross-sectional studies with adults and children in America who were asked to solve simple arithmetic tasks, suggest that it is best not to assume that adults' and children's brains exhibit similar activations when they are asked to perform similar tasks. Studies with adults found that retrieval versus procedural counting activated different parts of the brain (e.g., Grabner et al. 2009). Other studies that explored brain responses of adults who learned how to solve multi-digit arithmetic problems showed that increased proficiency with recently learned arithmetic facts showed reduced activity in certain parts of the brain and increased activity in other parts of the brain (e.g., Ischebeck et al. 2007). Cross-sectional neuroimaging studies show that children rely on different parts of the brain to solve arithmetic problems of the canonical form $a + b = ?$, where a and b are single digits. Although both groups may choose to use the same strategies to solve the same problems, the strategies may be more effortful for children than for adults (Rivera et al. 2005).

Research shows that young children use one of four strategies to add canonical number sentences such as $8 + 5 = ?$, moving from the more effortful strategy of counting procedures, such as counting fingers, to verbal counting, to the more efficient memory-based retrieval of number facts or decomposition (e.g., 8 + 2 + 3). It cannot be assumed that learning in adults or contrasts between children and adults is comparable to learning in the developing brain (Karmiloff-Smith 2010).

Because almost nothing is known about brain changes that accompany the transition from the use of procedural counting to direct retrieval, which is a critical aspect of children's early arithmetic development, Cho et al. (2011) examined how 103 children in northern California (age range from 7.0 to 9.9 years, 54 girls) used strategies to solve canonical addition problems. This choice of participants was strategic because children within this narrow age group begin to *develop* efficient problem solving strategies, in this case the use of retrieval strategies to solve canonical arithmetic sentences. Their neuroimaging studies demonstrated that retrieval and counting strategies during the developmental phase of learning are characterized by distinct patterns of activity in a distributed network of brain regions involved in arithmetic problem solving and controlled retrieval of arithmetic facts. The findings suggest that the reorganization and refinement of neural activity patterns in multiple brain regions plays a dominant role in the transition to memory-based arithmetic problem solving. Cho et al.'s work emphasizes that developmental changes cannot be inferred from, or characterized by, a gross comparison between adults and children or by examining the effects of training on novel problems in adults.

The plastic nature of the brain offers pedagogical opportunities whereby teachers can teach for the development of memory-based knowledge. Teaching children to use efficient strategies to perform mental operations increases performance of both normal performing children and those who may have difficulties recalling number facts. When children are able to perform arithmetic operations accurately and with automaticity, they have greater opportunities to perform more complex mathematical tasks (Menon 2010), such as exploring relationships between numbers and identifying patterns underpinning number sequences. Such developmental work with arithmetic cautions against extrapolating from Lee et al.'s (2007, 2010) work with proficient adults to children. These studies do, nonetheless, suggest some very interesting and important questions. How do children's brains (age 9–10 years) respond when they first learn to use the model method to represent quantitative information of canonical arithmetic expressions? Will their brains respond differently to the same sets of problems when they can use the model method competently? How will the brains of these two groups of children (the beginners and the competent users) respond when they first learn to use the model method to solve algebraic word problems and again when they use formal algebra to solve the same sets of problems?

2.5   Concluding Remarks

We have provided in this topical survey chapter some examples of the existing research in the learning and teaching of early algebra, as well as a sketch of the history and evolution of this field. We have described the current state of research with respect to the nature of early algebraic thinking, its processes, and its mathematical content areas. This research has yielded insights into approaches for developing students' early algebraic thinking, but has also signalled the need for teacher support in this area. It has produced theoretical frames for characterizing algebraic thinking, pattern generalization, and functional thinking, but more remains to be done, especially with respect to theorizing the algebraic aspects of students' work with number, operations, and properties. The development and use of digital tool environments that can encourage the growth of early algebraic thinking in the young learner is another underdeveloped area in this field. Current research has also alerted us, by means of its neurocognitive studies, to the different attentional demands made by the diagrammatic model method versus the alphanumeric algebraic method of problem representation and solution. We close with the comment that the field of early algebra research is a relatively young one, a field where new and exciting work is currently taking place and which offers the promise of equally interesting and informative results in the years to come.

Chapter 3
Summary and Looking Ahead

This topical survey, with its sketch of the history and evolution of the field of early algebra, has highlighted research related to the nature of early algebra, its learning, and its teaching. We have noted, in particular, that early algebraic thinking does not develop on its own without appropriate instructional support. And so, as we look ahead to the future, we recommend that further research be carried out in the following areas:

- The nature of classroom culture and the role of the teacher in fostering early algebraic reasoning.
- The forms of curricular activity that support early algebraic thinking.
- The nature of professional development that supports teachers' capacity to foster early algebraic thinking in the classroom.
- Theorizing about the study of number, operations, and properties in the context of early algebra.
- The use of neuroimaging techniques to inform the learning and teaching of early algebra.
- The development and use of digital tools to facilitate the teaching and learning of early algebra.
- The impact of early algebraic thinking on students' later study of algebra.

© The Author(s) 2016
C. Kieran et al., *Early Algebra*, ICME-13 Topical Surveys,
DOI 10.1007/978-3-319-32258-2_3

References

Ainley, J. (1999). Doing algebra type stuff: Emergent algebra in the primary school. In O. Zaslavsky (Ed.), *Proceedings of the 23rd Conference of the International Group for the Psychology of Mathematic Education* (Vol. 2, pp. 9–16). Haifa, Israel: PME.

Ainley, J. (2001). Research forum on early algebra. In M. van den Heuvel-Panhuizen (Ed.), *Proceedings of the 25th Conference of the International Group for the Psychology of Mathematics Education* (Vol. 1, pp. 129–159). Utrecht, The Netherlands: PME.

Ainley, J., Nardi, E., & Pratt, D. (1998). Constructing meaning for formal notation in Active Graphing. In I. Schwank (Ed.), *Proceedings of the First Conference of the European Society for Research in Mathematics Education* (Vol. I, pp. 189–200). Osnabrück, Germany: CERME.

Bass, H. B., & Ball, D. L. (2003). Foreword. In T. P. Carpenter, M. L. Franke, & L. Levi, *Thinking mathematically: Integrating arithmetic and algebra in elementary school* (pp. v–vii). Portsmouth, NH: Heinemann.

Blanton, M. L. (2008). *Algebra in elementary classrooms: Transforming thinking, transforming practice.* Portsmouth, NH: Heinemann.

Blanton, M., Brizuela, B. M., Gardiner, A. M., Sawrey, K., & Newman-Owens, A. (2015a). A learning trajectory in six-year-olds' thinking about generalizing functional relationships. *Journal for Research in Mathematics Education,46*, 511–558.

Blanton, M. L., & Kaput, J. J. (2004). Elementary grade students' capacity for functional thinking. In M. J. Høines & A. B. Fuglestad (Eds.), *Proceedings of the 28th Conference of the International Group for the Psychology of Mathematics Education* (Vol. 2, pp. 135–142). Bergen, Norway: PME.

Blanton, M. L., & Kaput, J. J. (2008). Building district capacity for teacher development in algebraic reasoning. In J. J. Kaput, D. W. Carraher, & M. L. Blanton (Eds.), *Algebra in the early grades* (pp. 361–388). New York: Routledge.

Blanton, M., Levi, L., Crites, T., & Dougherty, B. (2011). Developing essential understanding of algebraic thinking for teaching mathematics in grades 3–5. In B. J. Dougherty & R. M. Zbiek (Eds.), *Essential understandings series*. National Council of Teachers of Mathematics: Reston, VA.

Blanton, M., Stephens, A., Knuth, E., Gardiner, A. M., Isler, I., & Kim, J.-S. (2015b). The development of children's algebraic thinking: The impact of a comprehensive early algebra intervention in third grade. *Journal for Research in Mathematics Education,46*, 39–87.

Bolea, P., Bosch, M., & Gascón, J. (1998). The role of algebraization in the study of a mathematical organization. In I. Schwank (Ed.), *Proceedings of the First Conference of the European Society for Research in Mathematics Education* (Vol. II, pp. 135–145). Osnabrück, Germany: CERME.

Booth, L. R. (1984). *Algebra: Children's strategies and errors.* Windsor, UK: NFER-Nelson.

Bourke, S., & Stacey, K. (1988). Assessing problem solving in mathematics: Some variables related to student performance. *Australian Educational Researcher,15*, 77–83.

Britt, M. S., & Irwin, K. C. (2011). Algebraic thinking with and without algebraic representation: A pathway for learning. In J. Cai & E. Knuth (Eds.), *Early algebraization* (pp. 137–159). New York: Springer.

Brizuela, B., & Schliemann, A. (2004). Ten-year-old students solving linear equations. *For the Learning of Mathematics,24*(2), 33–40.

Cai, J. (2004). Developing algebraic thinking in the earlier grades: A case study of the Chinese elementary school curriculum. *The Mathematics Educator,8*(1), 107–130.

Cai, J., & Knuth, E. (Eds.). (2011). *Early algebraization*. New York: Springer.

Cai, J., Ng, S. F., & Moyer, J. (2011). Developing students' algebraic thinking in earlier grades: Lessons from China and Singapore. In J. Cai & E. Knuth (Eds.), *Early algebraization* (pp. 25–42). New York: Springer.

Carpenter, T. P., Fennema, E., Franke, M. L., Levi, L., & Empson, S. B. (1999). *Children's mathematics: Cognitively guided instruction*. Portsmouth, NH: Heinemann; Reston, VA: National Council of Teachers of Mathematics.

Carpenter, T. P., Franke, M. L., & Levi, L. (2003). *Thinking mathematically: Integrating arithmetic and algebra in elementary school*. Portsmouth, NH: Heinemann.

Carraher, D. W., & Schliemann, A. D. (2007). Early algebra and algebraic reasoning. In F. K. Lester Jr (Ed.), *Second handbook of research on mathematics teaching and learning* (pp. 669–705). Charlotte, NC: Information Age.

Carraher, D. W., & Schliemann, A. D. (2015). Powerful ideas in elementary school mathematics. In L. D. English & D. Kirshner (Eds.), *Handbook of international research in mathematics education* (3rd ed., pp. 191–218). New York: Taylor & Francis.

Carraher, D., Schliemann, A. D., & Brizuela, B. M. (2001). Can young students operate on unknowns? In M. van den Heuvel-Panhuizen (Ed.), *Proceedings of the 25th Conference of the International Group for the Psychology of Mathematics Education* (Vol. 1, pp. 130–140). Utrecht, The Netherlands: PME.

Carraher, D. W., Schliemann, A. D., Brizuela, B. M., & Earnest, D. (2006). Arithmetic and algebra in early mathematics education. *Journal for Research in Mathematics Education,37*, 87–115.

Cho, S., Ryali, S., Geary, D. C., & Menon, V. (2011). How does a child solve 7 + 8? Decoding brain activity patterns associated with counting and retrieval strategies. *Developmental Science,14*(5), 989–1001.

Collis, K. F. (1975). *A study of concrete and formal operations in school mathematics: A Piagetian viewpoint*. Melbourne, Australia: Australian Council for Educational Research.

Cooper, T. J., & Warren, E. (2011). Years 2 to 6 students' ability to generalise: Models, representations and theory for teaching and learning. In J. Cai & E. Knuth (Eds.), *Early algebraization* (pp. 187–214). New York: Springer.

Curriculum Planning & Development Division. (2006). *2006 Mathematics syllabus: Primary*. Singapore: Ministry of Education.

Cusi, A., & Malara, N. A. (2013). A theoretical construct to analyze the teacher's role during introductory activities to algebraic modelling. In B. Ubuz, Ç. Haser, & M. A. Mariotti (Eds.), *Proceedings of the 8th Congress of the European Society for Research in Mathematics Education* (pp. 3015–3024). Antalya, Turkey: CERME.

Cusi, A., Malara, N., & Navarra, G. (2011). Theoretical issues and educational strategies for encouraging teachers to promote a linguistic and metacognitive approach to early algebra. In J. Cai & E. Knuth (Eds.), *Early algebraization* (pp. 483–510). New York: Springer.

Davis, R. B. (1964). *Discovery in mathematics: A text for teachers*. Palo Alto, CA: Addison-Wesley.

Davis, R. B. (1975). Cognitive processes involved in solving simple algebraic equations. *Journal of Mathematical Behavior,1*(3), 7–35.

Davis, R. B. (1985). ICME-5 report: Algebraic thinking in the early grades. *Journal of Mathematical Behavior,4*, 195–208.

Davis, R. B. (1995). Why are they changing school algebra and who's doing it? *Journal of Mathematical Behavior,14*, 1–3.

Davydov, V. V., Gorbov, S. F., Mikulina, G. G., Savaleva, O. V. (1999). *Mathematics class 1* (edited by J. Schmittau). Binghamton: State University of New York.

Dougherty, B. J. (2003). Voyaging from theory to practice in learning: Measure Up. In N. A. Pateman, B. J. Dougherty, & J. T. Zilliox (Eds.), *Proceedings of the 27th Conference of the International Group for the Psychology of Mathematics Education* (Vol. 1, pp. 17–23). Honolulu: PME.

Dougherty, B. (2008). Measure Up: A quantitative view of early algebra. In J. J. Kaput, D. W. Carraher, & M. L. Blanton (Eds.), *Algebra in the early grades* (pp. 389–412). New York: Routledge.

Dougherty, B. J., & Slovin, H. (2004). Generalized diagrams as a tool for young children's problem solving. In M. J. Høines & A. B. Fuglestad (Eds.), *Proceedings of the 28th Conference of the International Group for the Psychology of Mathematics Education* (Vol. 2, pp. 295–302). Bergen, Norway: PME.

Driscoll, M. (1999). *Fostering algebraic thinking: A guide for teachers, grades 6–10*. Portsmouth, NH: Heinemann.

Ellis, A. B. (2007). The influence of reasoning with emergent quantities on students' generalizations. *Cognition and Instruction, 25*(4), 439–478.

English, L. D., & Warren, E. A. (1998). Introducing the variable through pattern exploration. *The Mathematics Teacher, 91*(2), 166–170.

Franke, M., Carpenter, T. P., & Battey, D. (2008). In J. J. Kaput, D. W. Carraher, & M. L. Blanton (Eds.), *Algebra in the early grades* (pp. 333–360). New York: Routledge.

Fujii, T. (2003). Probing students' understanding of variables through cognitive conflict problems: Is the concept of variable so difficult for students to understand? In N. A. Pateman, B. J. Dougherty, & J. T. Zilliox (Eds.), *Proceedings of the 27th Conference of the International Group for the Psychology of Mathematics Education* (Vol. 1, pp. 49–65). Honolulu, HI: PME.

Fujii, T., & Stephens, M. (2001). Fostering an understanding of algebraic generalisation through numerical expressions: The role of quasi-variables. In H. Chick, K. Stacey, J. Vincent, & J. Vincent (Eds.), *Proceedings of the 12th ICMI Study Conference: The Future of the Teaching and Learning of Algebra* (pp. 258–264). Melbourne, Australia: The University of Melbourne.

Goldenberg, E. P., & Shteingold, N. (2008). Early algebra: The Math Workshop perspective. In J. J. Kaput, D. W. Carraher, & M. L. Blanton (Eds.), *Algebra in the early grades* (pp. 449–478). New York: Routledge.

Grabner, R., Ansari, D., Koschutnig, K., Reishofer, G., Ebner, F., & Neuper, C. (2009). To retrieve or to calculate? Left angular gyrus mediates the retrieval of arithmetic facts during problem solving. *Neuropsychologia, 47*, 604–608.

Grouws, D. A. (Ed.). (1992). *Handbook of research on mathematics teaching and learning*. New York: Macmillan.

Hewitt, D. (2014). A symbolic dance: The interplay between movement, notation, and mathematics on a journey toward solving equations. *Mathematical Thinking and Learning, 16*, 1–31.

Ischebeck, A., Zamarian, L., Egger, K., Schocke, M., & Delazer, M. (2007). Imaging early practice effects in arithmetic. *NeuroImage, 36*, 993–1003.

Kail, R., & Park, Y. S. (1994). Processing time, articulation time, and memory span. *Journal of Experimental Child Psychology, 57*(2), 281–291.

Kaput, J. J. (1998). Transforming algebra from an engine of inequity to an engine of mathematical power by "algebrafying" the K-12 curriculum. In *The nature and role of algebra in the K-14 curriculum* (Proceedings of a National Symposium, 1997, organized by the National Council of Teachers of Mathematics, the Mathematical Sciences Education Board, and the National Research Council, pp. 25–26). Washington, DC: National Academy Press.

Kaput, J. J. (2008). What is algebra? What is algebraic reasoning? In J. J. Kaput, D. W. Carraher, & M. L. Blanton (Eds.), *Algebra in the early grades* (pp. 5–17). New York: Routledge.

Kaput, J.J., & Blanton, M. (2001). Algebrafying the elementary mathematics experience. Part 1: Transforming task structures. In H. Chick, K. Stacey, J. Vincent, & J. Vincent (Eds.), *Proceedings of the 12th ICMI Study Conference: The Future of the Teaching and Learning of Algebra* (pp. 344–351). Melbourne, Australia: The University of Melbourne.

Kaput, J. J., Blanton, M. L., & Moreno, L. (2008a). Algebra from a symbolization point of view. In J. J. Kaput, D. W. Carraher, & M. L. Blanton (Eds.), *Algebra in the early grades* (pp. 19–55). New York: Routledge.

Kaput, J. J., Carraher, D. W., & Blanton, M. L. (Eds.). (2008b). *Algebra in the early grades*. New York: Routledge.

Karmiloff-Smith, A. (2010). Neuroimaging of the developing brain: taking 'developing' seriously. *Human Brain Mapping,31*, 934–941.

Kazemi, E., & Stipek, D. (2001). Promoting conceptual thinking in four upper-elementary mathematics classrooms. *Elementary School Journal,102*, 59–80.

Khng, K. H., & Lee, K. (2009). Inhibiting interference from prior knowledge: Arithmetic intrusions in algebra word problem solving. *Learning and Individual Differences,19*, 262–268.

Kieran, C. (1981). Concepts associated with the equality symbol. *Educational Studies in Mathematics,12*, 317–326.

Kieran, C. (1989). The early learning of algebra: A structural perspective. In S. Wagner & C. Kieran (Eds.), *Research issues in the learning and teaching of algebra* (pp. 33–53). Reston, Virginia: NCTM.

Kieran, C. (1992). The learning and teaching of school algebra. In D. A. Grouws (Ed.), *Handbook of research on mathematics teaching and learning* (pp. 390–419). New York: Macmillan.

Kieran, C. (1996). The changing face of school algebra. In C. Alsina, J. Alvarez, B. Hodgson, C. Laborde, & A. Pérez (Eds.), *Eighth International Congress on Mathematical Education: Selected lectures* (pp. 271–290). Seville: S.A.E.M. Thales.

Kieran, C. (1997). Mathematical concepts at the secondary school level: The learning of algebra and functions. In T. Nunes & P. Bryant (Eds.), *Learning and teaching mathematics: An international perspective* (pp. 133–158). East Sussex, UK: Psychology Press.

Kieran, C. (2004). Algebraic thinking in the early grades: What is it? *The Mathematics Educator,8* (1), 139–151.

Kieran, C. (2011). Overall commentary on early algebraization: Perspectives for research and teaching. In J. Cai & E. Knuth (Eds.), *Early algebraization* (pp. 579–593). New York: Springer.

Küchemann, D. (1981). Algebra. In K. Hart (Ed.), *Children's understanding of mathematics* (pp. 11–16). London: John Murray.

Kwon, H., Reiss, A. L., & Menon, V. (2002). Neural basis of protracted developmental changes in visual-spatial working memory. *Proceedings of the National Academy of Sciences of the United States of America,99*(20), 13336–13341.

Lannin, J. K., Barker, D. D., & Townsend, B. E. (2006). Recursive and explicit rules: How can we build student algebraic understanding? *Journal of Mathematical Behavior,25*(4), 299–317.

Lee, K., Lim, Z. Y., Yeong, S. H. M., Ng, S. F., Venkatraman, V., & Chee, M. W. L. (2007). Strategic differences in algebraic problem solving: Neuroanatomical correlates. *Brain Research,1155*, 163–171.

Lee, K., Yeong, S. H. M., Ng, S. F., Venkatraman, V., Graham, S., & Chee, M. W. L. (2010). Computing solutions to algebraic problems using a symbolic versus a schematic strategy. *ZDM —The International Journal on Mathematics Education,42*, 591–605.

Linchevski, L. (1995). Algebra with numbers and arithmetic with letters: A definition of pre-algebra. *Journal of Mathematical Behavior,14*, 113–120.

Lins, R., & Kaput, J. J. (2004). The early development of algebraic reasoning: The current state of the field. In K. Stacey, H. Chick, & M. Kendal (Eds.), *The future of the teaching and learning of algebra: the 12th ICMI Study* (pp. 47–70). Dordrecht, The Netherlands: Kluwer Academic.

MacGregor, M., & Stacey, K. (1999). Learning the algebraic method of solving problems. *Journal of Mathematical Behavior,18*(2), 149–167.

Malara, N. A., & Navarra, G. (2003). *ArAl Project: Arithmetic pathways towards favouring pre-algebraic thinking*. Bologna, Italy: Pitagora.

Mason, J. (1996). Expressing generality and roots of algebra. In N. Bednarz, C. Kieran, & L. Lee (Eds.), *Approaches to algebra* (pp. 65–86). Dordrecht: Kluwer.

Mason, J. (with Graham, A. & Johnston-Wilder, S.). (2005). *Developing thinking in algebra*. London: Sage.

Matthews, P., Rittle-Johnson, B., McEldoon, K., & Taylor, R. (2012). Measure for measure: What combining diverse measures reveals about children's understanding of the equal sign as an indicator of mathematical equality. *Journal for Research in Mathematics Education,43*, 316–350.

Mavrikis, M., Noss, R., Hoyles, C., & Geraniou, E. (2013). Sowing the seeds of algebraic generalisation: Designing epistemic affordances for an intelligent microworld. In R. Noss & A. DiSessa (Eds.), *Journal of Computer Assisted Learning, 29*(1), 68–85.

Menon, V. (2010). Developmental cognitive neuroscience of arithmetic: implications for learning and education. *ZDM—The International Journal on Mathematics Education,42*, 515–525.

Moss, J., & London McNab, S. (2011). An approach to geometric and numeric patterning that fosters second grade students' reasoning and generalizing about functions and co-variation. In J. Cai & E. Knuth (Eds.), *Early algebraization* (pp. 277–301). New York: Springer.

Moyer, J., Huinker, D., & Cai, J. (2004). Developing algebraic thinking in the earlier grades: A case study of the U.S. *Investigations* curriculum. *The Mathematics Educator,8*(1), 6–38.

National Council of Teachers of Mathematics. (1989). *Curriculum and evaluation standards for school mathematics*. Reston, VA: NCTM.

National Council of Teachers of Mathematics. (2000). *Principles and standards for school mathematics*. Reston, VA: NCTM.

Ng, S. F. (2004). Developing algebraic thinking in early grades: Case study of the Singapore primary mathematics curriculum. *The Mathematics Educator,8*(1), 39–59.

Ng, S. F. (2012). *A theoretical framework for understanding the different attention resource demands of letter-symbolic versus model method*. Paper presented at Topic Study Group 9 of the 12th International Congress on Mathematical Education, Seoul, Korea. Available at: http://www.icme12.org/upload/UpFile2/TSG/0748.pdf

Ng, S. F., & Lee, K. (2009). The model method: Singapore children's tool for representing and solving algebraic word problems. *Journal for Research in Mathematics Education,40*, 282–313.

Noble, T., Nemirovsky, R., Wright, T., & Tierney, C. (2001). Experiencing change: The mathematics of change in multiple environments. *Journal for Research in Mathematics Education,32*, 85–108.

Orton, A. (Ed.). (1999). *Pattern in the teaching and learning of mathematics*. London: Cassell.

Radford, L. (2000). Signs and meanings in students' emergent algebraic thinking: A semiotic analysis. *Educational Studies in Mathematics,42*, 237–268.

Radford, L. (2003). Gestures, speech, and the sprouting of signs: A semiotic-cultural approach to students' types of generalizations. *Mathematical Thinking and Learning,5*, 37–70.

Radford, L. (2006). Algebraic thinking and the generalization of patterns: A semiotic perspective. In S. Alatorre, J. L. Cortina, M. Sáiz, & A. Méndez (Eds.), *Proceedings of the 28th annual meeting of the North American Chapter of the International Group for the Psychology of Mathematics Education* (Vol. 1, pp. 2–21). Mérida, MX: PME-NA.

Radford, L. (2010). Layers of generality and types of generalization in pattern activities. *PNA,4*(2), 37–62.

Radford, L. (2011a). Embodiment, perception and symbols in the development of early algebraic thinking. In B. Ubuz (Ed.), *Proceedings of the 35th Conference of the International Group for the Psychology of Mathematics Education* (Vol. 4, pp. 17–24). Ankara, Turkey: PME.

Radford, L. (2011b). Grade 2 students' non-symbolic algebraic thinking. In J. Cai & E. Knuth (Eds.), *Early algebraization* (pp. 303–322). New York: Springer.

Radford, L. (2014). The progressive development of early embodied algebraic thinking. *Mathematics Education Research Journal,26*, 257–277.

Radford, L., & Roth, W.-M. (2011). Intercorporeality and ethical commitment: An activity perspective on classroom interaction. *Educational Studies in Mathematics,77*, 227–245.

Rivera, F. (2013). *Teaching and learning patterns in school mathematics: Psychological and pedagogical considerations.* New York: Springer.

Rivera, S. M., Reiss, A. L., Eckert, M. A., & Menon, V. (2005). Developmental changes in mental arithmetic: evidence for increased functional specialization in the left inferior parietal cortex. *Cerebral Cortex,15*, 1779–1790.

Rojano, T., & Sutherland, R. (2001). Arithmetic world—algebra world. In H. Chick, K. Stacey, J. Vincent, & J. Vincent (Eds.), *Proceedings of the 12th ICMI Study Conference: The Future of the Teaching and Learning of Algebra* (pp. 515–522). Melbourne, Australia: The University of Melbourne.

Roschelle, J., Shechtman, N., Tatar, D., Hegedus, S., Hopkins, B., Empson, S., & Gallagher, L. P. (2010). Integration of technology, curriculum, and professional development for advancing middle school mathematics: Three large-scale studies. *American Educational Research Journal,47*(4), 833–878.

Russell, S. J. (2015). *Mathematical argument and productive lingering.* Teachers Development Group Leadership Seminar, Portland, OR, March 2015. Available from: http://www. teachersdg.org/2015%20Seminar%20Docs/OK_SJRussell_Lingering_Argument.pdf

Russell, S. J., Schifter, D., & Bastable, V. (2011). *Connecting arithmetic to algebra.* Portsmouth, NH: Heinemann.

Russell, S. J., Schifter, D., Bastable, V., & Franke, M. (submitted). *Bringing early algebra to the elementary classroom: Results of a professional development program for teachers.*

Russell, S. J., Schifter, D., Bastable, V., Higgins, T., & Kasman, R. (in press). *Mathematical argument in the elementary classroom: A yearlong focus on the arithmetic operations.* Portsmouth, NH: Heinemann.

Schifter, D. (1999). Reasoning about operations: Early algebraic thinking in grades K-6. In L. V. Stiff & F. R. Curcio (Eds.), *Developing mathematical reasoning in grades K-12* (pp. 62–81). Reston, VA: National Council of Teachers of Mathematics.

Schifter, D., Bastable, V., Russell, S. J., Riddle, M., & Seyferth, L. (2008a). Algebra in the K-5 classroom: Learning opportunities for students and teachers. In C. E. Greenes & R. Rubenstein (Eds.), *Algebra and algebraic thinking in school mathematics. 2008 Yearbook of the National Council of Teachers of Mathematics* (pp. 263–277). Reston, VA: NCTM.

Schifter, D., Monk, S., Russell, S. J., & Bastable, V. (2008b). What does understanding the laws of arithmetic mean in the elementary grades? In J. J. Kaput, D. W. Carraher, & M. L. Blanton (Eds.), *Algebra in the early grades* (pp. 413–448). New York: Routledge.

Schliemann, A., Carraher, D. W., Brizuela, B., Earnest, D., Goodrow, A., Lara-Roth, S., & Peled, I. (2003). Algebra in elementary school. In N. A. Pateman, B. J. Dougherty, & J. T. Zilliox (Eds.), *Proceedings of the 27th Conference of the International Group for the Psychology of Mathematics Education* (Vol. 4, pp. 127–134). Honolulu, HI: PME.

Schmittau, J., & Morris, A. (2004). The development of algebra in the elementary mathematics curriculum of V. V. Davydov. *The Mathematics Educator, 8*(1), 60-87.

Sfard, A. (1991). On the dual nature of mathematical conceptions: Reflections on processes and objects as different sides of the same coin. *Educational Studies in Mathematics,22*, 1–36.

Sfard, A., & Linchevski, L. (1994). The gains and the pitfalls of reification: The case of algebra. *Educational Studies in Mathematics,26*(2/3), 191–228.

Slavit, D. (1999). The role of operation sense in transitions from arithmetic to algebraic thought. *Educational Studies in Mathematics,37*, 251–274.

Stacey, K. (1989). Finding and using patterns in linear generalizing problems. *Educational Studies in Mathematics,20*, 147–164.

Stacey, K., Chick, H., & Kendal, M. (Eds.). (2004). *The future of the teaching and learning of algebra: the 12th ICMI Study.* Dordrecht, The Netherlands: Kluwer Academic.

Stacey, K., & MacGregor, M. (1997). Ideas about symbolism that students bring to algebra. *The Mathematics Teacher,90*(2), 110–113.

Stein, M. K., Remillard, J., & Smith, M. S. (2007). How curriculum influences student learning. In F. K. Lester Jr (Ed.), *Second handbook of research on mathematics teaching and learning* (pp. 319–369). Greenwich, CT: Information Age Publishing.

Steinweg, A. S. (2001). Children's understanding of number patterns. In M. van den Heuvel-Panhuizen (Ed.), *Proceedings of the 25th Conference of the International Group for the Psychology of Mathematics Education* (Vol. 1, pp. 203–206). Utrecht, The Netherlands: PME.

Sutherland, R. (1993). Connecting theory and practice: Results from the teaching of Logo. *Educational Studies in Mathematics,24*, 95–113.

TERC. (1998). *Investigations in number, data, and space*. Menlo Park, CA: Dale Seymour Publications.

Usiskin, Z. (1988). Conceptions of school algebra and uses of variables. In A. Coxford (Ed.), *Ideas of algebra: K-12* (pp. 8–19). Reston, VA: NCTM.

van Ameron, B. A. (2002). *Reinvention of early algebra—Developmental research on the transition from arithmetic to algebra* (doctoral thesis). Utrecht University, Freudenthal Institute. Available from: http://dspace.library.uu.nl/bitstream/handle/1874/874/full.pdf?sequence=18

Wagner, S., & Kieran, C. (Eds.). (1989). *Research issues in the learning and teaching of algebra. Vol. 4. Research agenda for mathematics education*. Reston, VA: National Council of Teachers of Mathematics.

Warren, E. (2002). Unknowns, arithmetic to algebra: two exemplars. In A. D. Cockburn & E. Nardi (Eds.), *Proceedings of the 26th Conference of the International Group for the Psychology of Mathematics Education* (Vol. 4, pp. 361–368). Norwich, UK: PME.

Warren, E. (2003). Young children's understanding of equals: A longitudinal study. In N. A. Pateman, B. J. Dougherty, & J. T. Zilliox (Eds.), *Proceedings of the 27th Conference of the International Group for the Psychology of Mathematics Education* (Vol. 4, pp. 379–386). Honolulu, HI: PME.

Warren, E., & Cooper, T. J. (2008). Patterns that support early algebraic thinking in the elementary school. In C. E. Greenes & R. Rubenstein (Eds.), *Algebra and algebraic thinking in school mathematics. 70th Yearbook of the National Council of Teachers of Mathematics* (pp. 113–126). Reston, VA: NCTM.

Warren, E., & Cooper, T. (2009). Developing mathematics understanding and abstraction: The case of equivalence in the elementary years. *Mathematics Education Research Journal,21*(2), 76–95.

Watanabe, T. (2011). A critical foundation for school algebra in Japanese elementary school mathematics. In J. Cai & E. Knuth (Eds.), *Early algebraization* (pp. 109–124). New York: Springer.

Webb, N., Franke, M., Ing, M., Wong, H., Fernandez, C., Shin, N., & Turrou, A. (2014). Engaging with other's mathematical ideas: Interrelationships among student participation, teachers' instructional practice, and learning. *International Journal for Educational Research,63*, 79–93.

Zazkis, R., & Liljedahl, P. (2002). Generalization of patterns: the tension between algebraic thinking and algebraic notation. *Educational Studies in Mathematics,49*, 379–402.

Further Reading

Blanton, M., Brizuela, B. M., & Stephens, A. C. (2016). *Elementary children's algebraic thinking*. Plenary paper prepared for presentation at Topic Study Group 10 of 13th International Congress on Mathematical Education (ICME13).

Carraher, D. W., & Schliemann, A. D. (2016). *Functional relations in early algebraic thinking*. Plenary paper prepared for presentation at Topic Study Group 10 of 13th International Congress on Mathematical Education (ICME13).

Malara, N. A., & Navarra, G. (2016). *Epistemological issues in early algebra: Offering teachers new words and paradigms to promote pupils' algebraic thinking.* Plenary paper prepared for presentation at Topic Study Group 10 of 13th International Congress on Mathematical Education (ICME13).

Mason, J. (2016). *How early is too early for thinking algebraically?* Plenary paper prepared for presentation at Topic Study Group 10 of 13th International Congress on Mathematical Education (ICME13).

GPSR Compliance
The European Union's (EU) General Product Safety Regulation (GPSR) is a set
of rules that requires consumer products to be safe and our obligations to
ensure this.

If you have any concerns about our products, you can contact us on

ProductSafety@springernature.com

In case Publisher is established outside the EU, the EU authorized
representative is:

Springer Nature Customer Service Center GmbH
Europaplatz 3
69115 Heidelberg, Germany